YOUR KNOWLEDGE HAS VALUE

- We will publish your bachelor's and
  master's thesis, essays and papers

- Your own eBook and book -
  sold worldwide in all relevant shops

- Earn money with each sale

Upload your text at www.GRIN.com
and publish for free

Bibliographic information published by the German National Library:

The German National Library lists this publication in the National Bibliography; detailed bibliographic data are available on the Internet at http://dnb.dnb.de .

This book is copyright material and must not be copied, reproduced, transferred, distributed, leased, licensed or publicly performed or used in any way except as specifically permitted in writing by the publishers, as allowed under the terms and conditions under which it was purchased or as strictly permitted by applicable copyright law. Any unauthorized distribution or use of this text may be a direct infringement of the author s and publisher s rights and those responsible may be liable in law accordingly.

Imprint:

Copyright © 2016 GRIN Verlag, Open Publishing GmbH
Print and binding: Books on Demand GmbH, Norderstedt Germany
ISBN: 9783668379749

This book at GRIN:

http://www.grin.com/en/e-book/351314/a-comparative-study-on-traditional-herbal-medicines-of-mannans-tribal

Prem Jose Vazhacharickal, Jiby John Mathew, Sajeshkumar N.K, Nijamol Varghese

A comparative study on traditional herbal medicines of Mannans (tribal group in Idukki, Kerala) for the treatment of diabetes

GRIN Publishing

GRIN - Your knowledge has value

Since its foundation in 1998, GRIN has specialized in publishing academic texts by students, college teachers and other academics as e-book and printed book. The website www.grin.com is an ideal platform for presenting term papers, final papers, scientific essays, dissertations and specialist books.

Visit us on the internet:

http://www.grin.com/

http://www.facebook.com/grincom

http://www.twitter.com/grin_com

A comparative study on traditional herbal medicines of Mannans (tribal group in Idukki, Kerala) for the treatment of diabetes

Prem Jose Vazhacharickal, Jiby John Mathew, Sajeshkumar N.K and Nijamol Varghese

ACKNOWLEDGEMENTS

Firstly we thank **God Almighty** whose blessing were always with us and helped us to complete this project work successfully.

We wish to thank our beloved Manager **Rev. Fr. Dr. George Njarakunnel,** Respected Principal **Dr. V.J.Joseph,** Vice Principal **Fr. Joseph Allencheril**, Bursar **Shaji Augustine** and the Management for providing all the necessary facilities in carrying out the study. We express our sincere thanks to **Mr. Binoy A Mulanthra** (lab in charge, Department of Biotechnology) for the support. This research work will not be possible with the co-operation of many farmers.

We are gratefully indebted to our teachers, parents, siblings and friends who were there always for helping us in this project.

Prem Jose Vazhacharickal*, Jiby John Mathew, Sajeshkumar N.K and Nijamol Varghese

*Address for correspondence
Assistant Professor
Department of Biotechnology
Mar Augusthinose College
Ramapuram-686576
Kerala, India
premjosev@gmail.com

Table of contents

Table of figures

Table of tables

List of abbreviations

AMPK	: AMP activated protein kinase
CDVD	: Chronic degenerative valve disease
Cu++	: Cupric
CVD	: Cardiovascular disease
DPP-IV	: Dipeptidyl peptidase-IV
GLP-1	: Glucon-like peptide 1
H_2SO_4	: Sulphuric acid
HCl	: Hydrochloric acid
IGF	: Insulin-like growth factor
IL	: Interleukin
LDL	: Low-density lipoprotein
MCP	: Monocyte chemoattractant protein
NaOH	: Sodium hydroxide
NH_3	: Ammonia
NH_4OH	: Ammonium hydroxide
OS	: Oxidative stress
PPARγ	: Peroxisome proliferator- activated receptor gamma
SPSS	: Statistical package for social science
T2D	: Type 2 diabetes
TNF-α	: Tumour necrosis factor-alpha
TZDs	: Thiazolidinedione's
WHO	: World health organization

A comparative study on traditional herbal medicines of Mannans (tribal group in Idukki, Kerala) for the treatment of diabetes

Prem Jose Vazhacharickal[1]*, Sajeshkumar N.K[1], Jiby John Mathew[1] and Nijamol Varghese[1]

* premjosev@gmailcom

[1]Department of Biotechnology, Mar Augusthinose College, Ramapuram, Kerala, India-686576

Abstract

Diabetes is a complex metabolic disorder resulting from either insulin insufficiency or insulin dysfunction. Diabetes mellitus affects most of the people in both developed and developing countries. The treatment of diabetes with synthetic drugs is costly and chances of side effects are high. Phytomedicine has been used since ancient times in various parts of the world where access to modern medicine is limited. Medicinal plants and phytoconstituents play an important role in the management of diabetes mellitus especially in developing countries where resources are meagre. Phytochemicals identified from medicinal plants present an exciting opportunity for the development of new types of therapeutics for diabetes mellitus. Most prevalent among phytochemical groups are the alkaloids, glycosides, polysaccharides, and phenolics such as flavonoids, terpenoids and steroids. This study aims to provide a comparative information about various plants used for antidiabetes treatment in a trial community (Mannans) and their constituents, which have been shown to display potent hypoglycemic activity. The current study was focused on four plants named *Tinospora cordifolia*, *Ensete superbum*, *Coccinia grandis* wild and *Apama siliquosa* Lam. which is locally known as Amruthu, Kalluvazha, Kattukoval, and Alpam respectively. The plant parts such as seed, leaves, stem and bark were used by the local people. The plant materials prepared as decoction, infusion, aqueous extracts in milk or honey were used for the treatment of diabetes. The plants that are being used by the local people of the study area have been isolated from plants for the treatment of diabetes. The efficacy of these ethnomedicinal plants needs to be subjected to pharmacological validation. it was found that *Coccinia grandis* was more efficient than the other plants. Some antidiabetic plants may exert their action by stimulating the function or number of beta-cells and thus increasing insulin release. The study highlighted the central role of traditional herbal medicine for the treatment of diabetes in tribal community. Ethnobotanical survey is most useful for scientists, research scholars and scientific companies for further studies on isolation and identification of active compounds that can be formulated into antidiabetic drugs.

Keywords: *Coccinia grandis*; ethnobotanical; diabetes; Tribal medicines; *Tinospora cordifolia*

1. Introduction

Diabetes is an endocrinological metabolic disorder characterized by chronic hyperglycemia, polyuria, polydipsia, polyphagia, emaciation and weakness due to disturbance in carbohydrate, fat and protein metabolism associated with absolute or relative deficiency in insulin secretion and/or insulin action1. Diabetes is becoming the third "killer" of the health of mankind along with cancer, cardiovascular and cerebrovascular disease (Donga et al., 2010). Synthetic antidiabetic agents like sulfonylureas, biguanides, glucosidase inhibitors and thiazolindiones are being expensive and produce serious side effects (Venkatesh et al., 2003). Further their use is not safe during pregnancy. Herbal therapy recommended for the treatment of diabetes throughout the world. Herbal drugs are prescribed widely because of their effectiveness, less side effects and relatively low cost (Venkatesh et al., 2003). According to the World Health Organization (WHO) more than one million people rely on herbal medicines to some extents. As an increase in demand by patients to use of natural products with anti-diabetic activity, investigations on hypoglycaemic agents derived from medicinal plants have gained popularity in recent years.

The World Health Organization defined diabetes mellitus as "a metabolic disorder of multiple etiologies characterized by chronic hyperglycaemia with disturbances in carbohydrate, fat and protein metabolism resulting from defects in insulin secretion, insulin action, or both" (WHO, 2013). Diabetes is a disorder of carbohydrate, fat and protein metabolism attributed to diminished production of insulin or mounting resistance to its action. Diabetes is a chronic disease that occurs either when the pancreas does not produce enough insulin or when the body cannot effectively use the insulin that it produces. Insulin is a hormone that regulates blood sugar. The increased production and ineffective scavenging of reactive oxygen species may play a critical role in diabetes mellitus. The disturbance of antioxidants defence system in diabetes is mainly because of alteration in antioxidant enzymes, impaired glutathione metabolism, and decreased ascorbic acid levels. It is becoming the third "killer" of the health of mankind along with cancer, cardiovascular and cerebrovascular diseases (Chauhan et al., 2010).

Hyperglycaemia, or raised blood sugar, is a common effect of uncontrolled diabetes and over the time leads to serious damages to many of the body's systems, especially the nerves and blood vessels. Over the time, diabetes can damage the heart, blood vessels, eyes, kidneys and nerves. Diabetes increases the risk of heart disease and stroke. 50% of people with diabetes die of cardiovascular disease. Combined with reduced blood flow, neuropathy (nerve damage) in the foot increases the chance of foot ulcers, infection and eventual need for limb amputation.

Plants have always been a good source of drugs. The ethnobotanical information reports about 800 plants that may possess anti-diabetic potential. The beneficial uses of medicinal plants in traditional system of medicine of many cultures are extensively documented. Several plants have been used as dietary adjuvant and in treating the number of diseases even without any knowledge on their proper functions and constituents. This practice may be attributed to the uncompromised cost and side effects of synthetic hypoglycaemic agent. Although numerous synthetic drugs were developed for the treatment of diabetes mellitus but the safe and effective treatment paradigm is yet to be achieved (Patel et al., 2012).

WHO has recommended the evaluation of traditional plant treatments for diabetes as they are effective, non-toxic, with less or no side effects and are considered to be excellent aspirants for oral therapy (Shokeen et al., 2008). A scientific validation of several plant species has proved the efficiency of the botanicals in reducing the sugar level. From the reports on their potential effectiveness against diabetes, it is assumed that the phytochemicals have a major role in the management of diabetes, which needs further exploration for the necessary development of drugs and nutraceuticals from natural resources.

Herbal treatments for diabetes have been used in patients with insulin-dependent and non-insulin-dependent diabetes, diabetic retinopathy, diabetic peripheral neuropathy, etc however many herbal remedies used today have not undergone careful scientific evaluation and some have the potential to cause serious toxic effects and major drug-to-drug interaction. Compounds with different structure but with the same therapeutic activity isolated from different plant species act as active moieties for the treatment of various diseases. The use of these plants and phyto-constituents may delay the development of diabetic complications and may regulate the metabolic abnormalities through a variety of mechanisms (Mukherjee et al., 2006). Moreover, during the past few years many phytochemicals responsible for anti-diabetic effects have been isolated from the plants. Several phytoconstituents such as alkaloids, glycosides, flavonoids, saponins, dietary fibres, polysaccharides, glycolipids, peptidoglycans, amino acids and others obtained from various plant sources that have been reported as potent hypoglycemic agents.

In spite of the advent of the modern medicines, tribal populations are still practicing the art of herbal medicine. The knowledge of the use of medicinal plants and their properties was transmitted from generation to generation (Subodh, 2010). The present study was conducted among a tribal community, Mannans, residing in Kumily Panchayth, Idukki district, Kerala state. The objectives of this study investigates the fundamental scientific bases for the use of some tribal medicinal plants for the treatment of diabetes by defining and enumerating the percentage of crude phytochemical constituents present in these plants and also finds out which plant was most effective one.

1.1 Taxonomical classification *Tinospora cordifolia* (plant 1)

Kingdom: Plantae-- planta, plantes, plants, vegetal

Subkingdom: Tracheobionta -- vascular plants

Division: Magnoliophyta -- angiosperms, flowering plants, phanerogames

Class: Magnoliopsida -- dicots, dicotyledones, dicotyledons

Order: Ranunculales

Family: Menispermaceae

Genus: Tinospora

Species: *Tinospora cordifolia*

1.2 Taxonomical classification *Ensete superbum* (plant 2)

Kingdom: Plantae-- planta, plantes, plants, vegetal

Subkingdom: Tracheobionta -- vascular plants

Division: Magnoliophyta -- angiosperms, flowering plants, phanerogames

Class: Liliopsida

Order: Zingiberales

Family: Musaceae

Genus: Ensete

Species: *Ensete superbum*

1.3 Taxonomical classification- *Coccinia grandis* (plant 3)

Kingdom: Plantae-- planta, plantes, plants, vegetal

Subkingdom: Tracheobionta -- vascular plants

Division: Magnoliophyta -- angiosperms, flowering plants, phanerogames

Class: Magnoliopsida -- dicots, dicotyledones, dicotyledons

Order: Violales

Family: Cucurbitaceae

Genus: Coccinia Wight

Species: *Coccinia grandis*

1.4 Taxonomical classification- *Apama siliquosa* Lam. (plant 4)

Kingdom: Plantae-- planta, plantes, plants, vegetal

Subkingdom: Tracheobionta -- vascular plants

Division: Magnoliophyta -- angiosperms, flowering plants, phanerogames

Class: Magnoliopsida -- dicots, dicotyledones, dicotyledons

Order: Piperales

Family: Aristolochiaceae

Genus: Thoetta

Species: *Apama siliquosa* Lam.

2. Review of literature

Diabetes is a chronic metabolic disorder characterized by high blood glucose levels. This is either as a result of insufficient endogenous insulin production by the pancreatic beta cells (otherwise known as type-1diabetes); or impaired insulin secretion and/or action (type-2diabetes). Type-1 diabetes's an autoimmune disease characterized by T-cell mediated destruction of the pancreatic beta cells. In type-2 diabetes, there is a gradual development of insulin resistance and beta cell dys function, strongly associated with obesity and a sedentary lifestyle (Zimmet et al.,

2001). Due to a higher incidence of the risk factors, the prevalence of diabetes is increasing worldwide, but more evidently in developing countries. Current estimates indicate a 69% increase in the number of adults that would be affected by the disease between 2010 and 2030, compared to 20% for developed countries (Shaw et al., 2010).

Administration of exogenous insulin is the treatment for all type-1 diabetic patients and for some type-2 patients who do not achieve adequate blood glucose control with oral hypoglycaemic drugs. Current drugs used in diabetes management can be categorized into three groups. Drugs in the first group increase endogenous insulin availability. These include the sulphonylureas such as glibenclamide, the glinides, insulin analogues, glucagon-like peptide1 (GLP-1) agonists and dipeptidyl peptidase-IV (DPP-IV) inhibitors. The first two members of this group act on the sulfonyl urea receptor in the pancreas to promote insulin secretion. GLP-1 agonists and DPP-IV inhibitors on the other hand act on the ileal cells of the small intestine. The second group of drugs enhance the sensitivity of insulin. This includes the thiazolidine-diones, which are agonists of the peroxisome proliferator- activated receptor gamma (PPARγ) and the biguanidemet formin. The third group comprises the α-glucosidase inhibitors such as acarbose, which reduce the digestion of polysaccharides and their bioavailability (Chehade and Mooradian, 2000; Sheehan, 2003).

All the existing therapies however have limited efficacy, limited tolerability and/or significant mechanism based side effects (Moller 2001; Rotenstein et al., 2012). Despite the existing pharmacotherapy, it is still difficult to attain adequate glycemic control amongst many diabetic patients due to the progressive decline in β-cell function (Wallace and Matthews, 2000). In Nigeria, polytherapy with two or more hypoglycaemic agents to achieve better glucose control is common practice (Yusuff et al., 2008). There is also a high incidence of diabetic complications and hyperglycaemic emergencies (Gill et al. 2009; Ogbera et al., 2007, 2009). In the presence of these, the number of prescribed drugs increases to an average of four per day for each patient (Enwere et al., 2006). This need for the chronic intake of a large number of drugs with their attendant side effects in addition to their high costs which is often borne by the patients themselves is the identified reason for non-adherence to therapy amongst diabetic patients. As a result, patients often have recourse to alternative forms of therapy such as herbal medicines (Yusuff et al., 2008).

A number of reviews on medicinal plants used in the management of diabetes in different parts of the world (Bailey and Day, 1989; Marles and Farnsworth, 1995), as well as those used specifically in certain regions, such as in West Africa (Bever, 1980), Central America (Andrade-Cetto and Heinrich, 2005) and Asia (Grover et al., 2002). These reviews have highlighted the dependence of a large percentage of the world population on traditional medicine for diabetes management. This is also corroborated by the WHO fact sheet (No.134), which estimates that about 80% of the population in African and Asian countries rely on traditional medicine for their primary health care (WHO, 2008). It also recognizes traditional medicine as 'an accessible, affordable and culturally acceptable form of health care trusted by large numbers of people, which stands out as a way of coping with the relentless rise of chronic non-communicable diseases in the midst of so a ring health-care costs and nearly universal austerity' (WHO, 2013). Ethnobotanical surveys of plants traditionally used in diabetes management in different parts of Nigeria have been carried out (Abo et al., 2008; Etukand Mohammed, 2009; Gbolade, 2009; Soladoye et al., 2012). These medicinal plants are used either alone as a primary therapeutic choice, or in conjunction with conventional medicines. On an average, approximately 50% of diabetic patients visiting hospitals in urban cities like Lagos and Benin have used some forms of traditional medicine during the course of their disease management (unpublished results of field work conducted by first author). Unfortunately, clinicians are either unaware of their patients 'herb use or the identity of the herbal product being taken. To complicate matters further, herbal practitioners are usually unwilling to divulge the identity of the constituents of their preparations to patients. Most patients are also not interested in finding this out as they consider herbal preparations to be 'safe'; there by making it difficult to ascertain if the herb may have a significant contributory role to the efficacy or failure of the treatment.

In a systematic review of herbs and supplements clinically used for glycemic control, *Allium sativum, Aloe Vera* and *Momordica charantia* were the only identified plants used in Nigeria. This inclusion was however based on clinical studies carried out outside Nigeria (Yeh et al., 2003). This indicates the lack of information about the clinical use of plants in diabetes management in Nigeria, despite widespread traditional use. In line with the increasing importance of traditional medicine in various health care systems around the world, the WHO Traditional Medicine Strategy has recently been updated. 'The goals of the strategy are to support

Member States in (a) harnessing the potential contribution of traditional medicine to health, wellness and people- centred health-care; and (b) promoting the safe and effective use of traditional medicine by regulating, researching and integrating traditional medicine products, practitioners and practice into health systems where appropriate' (WHO, 2013). Given that diabetes is now considered as one of the main threats to human health in the 21st century (Zimmet et al., 2001), there might be an even greater reliance by diabetic patients in Nigeria on herbal medicines used in its management. Unfortunately, pharmacological and toxicological evidences validating the safety and efficacy of these medicinal plants are not readily available. The objective of this paper is to collect as much as possible, available information about medicinal plants traditionally used in diabetes management in Nigeria. In doing so, we aim to promote the rational use of these plants based on pharmacological evidence for their therapeutic use and their toxic/interaction profile.

Medicinal plants are the main source of organic compounds such as polyphenols, tannins, alkaloids, carbohydrates, terpenoids, steroids and flavonoids. These organic compounds represent a source for the discovery and development of new types of anti-diabetic molecules. Many compounds isolated from plant sources have been reported to show anti diabetic activity. Phytochemicals are bioactive compounds contained in the plants that have no energy value. These compounds are usually produced in the plants to protect them against pests, diseases, to control the growth, or as pigments, essences and aromas (Perez Vizcaino et al., 2006). Currently, it is known a huge amount of phytochemicals which mainly consists of flavonoids, glucosinolates (isothiocyanates and indoles), phenolic acids, phytates, phytoestrogens (isoflavones and lignans), fats and oils contained in vegetables, fruits, cereals, legumes and other plant sources (Surh, 2002). They are characterized by various biological effects including anti-obesity, cholesterol lowering and anti- diabetic properties.

Diabetes affects about 5% of the global population and management of diabetes without any side effects is still a challenge to the medical system. In India the treatment of this disorder takes three main forms: (i) Diet and exercise (ii) Insulin replacement therapy and (iii) the use of oral hypoglycaemic agents. Currently available synthetic anti-diabetic agents like sulfonylurea's, biguanides, thiazolidinedione's (TZDs), a glycosidase inhibitors etc. besides being expensive produce serious side effects. Apart from currently available therapy, herbal medicines recommended for treatment of diabetes throughout the world. Herbal drugs are prescribed widely because of their effectiveness, less side effects and relatively low cost (Venkatesh et al., 2003). Numerous studies have confirmed the strong association between diet rich in plant foods and health the positive effects of these foods may rely on their content on phytochemicals, antioxidant, vitamins and fiber. Most of these dietary compounds contribute to a well redox balance by several mechanisms, such direct scavenging or neutralization of free radicals, modulation of enzyme activity and expression, and anti-inflammatory action. Phytochemicals are bioactive compounds contained in the plants that have the potential for offering protection against a range of non-communicable diseases like diabetes, cancer, cardiovascular disease and cataract.

Currently, it is known a huge amount of phytochemicals which mainly consists of flavonoids, glucosinolates (isothiocyanates and indoles), phenolic acids, phytates, phytoestrogens (isoflavones and lignans), fats and oils contained in vegetables, fruits, cereals, legumes and other plant sources (Surh, 2002). Phytochemicals like resveratrol, found in nuts and red wine, has antioxidant, antithrombotic, and anti-inflammatory properties, and inhibits carcinogenesis. Lycopene, a potent antioxidant carotenoid in tomatoes and other fruits, monoterpenes in citrus fruits is thought to protect against prostate and other cancers, and inhibits tumour cell growth in animals (Rao, 2003). Camphene is bicyclic monoterpens and present in essential oil of ginger (*Zingiber officianle*) and tulsi (*Ocimum sanctum*). Camphene having antioxidant, antimicrobial, chemo preventive (Lai and Roy, 2004) and anti lipemic activity (Tang et al., 2008).

Polyphenols constitute the most abundant phytochemicals provided by food of plant origin, being widely distributed in fruits, vegetables, whole cereals, coffee (*Coffea arabica*), cacao (*Theobroma cacao*), and tea (*Camellia sinensis*). In recent years numerous in vitro and animal studies have provided evidence that polyphenols may be protective against oxidative-triggered pathologies, including cardiovascular disease (CVD), metabolic disorders, cancer, and obesity. Polyphenols may have anti-obesity, anti-inflammatory, anti-diabetic, and anticancer properties through multiple mechanisms: they act by modulating inflammation and redox state, by regulating adipocyte differentiation and lipid metabolism, by inhibiting pancreatic lipase activity and intestinal permeability, and by interacting with gut micro biota. (Chattopadhyaya, 1996) According to their chemical structure, polyphenols are classified into different categories: phenolic acids, stilbenes, flavonoids (flavonols, flavanols, anthocyanins, flavanones, flavones, flavanonols, and isoflavones), chalcones, lignans and curcuminoids. Ferulic acid, a phenolic acid present in whole wheat, chocolate, apples, oranges, oregano, and sage, has been proved to be effective against high fat induced hyperlipidaemia and oxidative stress, via regulation of insulin secretion and regulation of antioxidant and lipogenic enzyme activities (Lacueva et al., 2011). Resveratrol, a primarily found in red grapes, apples, and peanuts, can be useful to counteract obesity, metabolic disorders, CVD, and cancer, through multiple actions: it increases mitochondrial activity, counteracts lipid accumulation, decreases inflammation, improves insulin signalling and modulates redox balance. Several mechanisms have been proposed for the hypoglycaemic effect of phytochemicals such as inhibition of carbohydrate metabolizing enzymes, β-cell regeneration and enhancing insulin releasing activity. Phytochemicals obtained from plant sources like catechin, ellagic acid, eugenol, kaempferol, berberin etc. have been reported to possess anti-diabetic activity (Son et al., 2011). Catechins are the most abundant flavonols contained in tea; they are also present in cocoa, grapes, and red wine. In streptozotoc in diabetic rats, establishes a hypoglycaemic condition paralleled by a better lipid profile. Epicatechin-enriched diet reduces insulin-like growth factor-1 (IGF-1) levels and prolongs lifespan in diabetic mice; similar results have also been found in *Drosophila melanogaster* (Engelhard, 2006). In humans, catechins have been proven to ameliorate blood pressure, low-density lipoprotein (LDL)-cholesterol, obesity, and chronic degenerative valve disease (CDVD) risk factors. Catechin-rich beverages (green tea containing about 600 mg catechins) improve obesity and glycaemia in type 2 diabetes patients (Ahmad, 1999). Daily supplementation of 379 mg green tea extracts reduces blood pressure, inflammatory biomarkers, and oxidative stress, and improves parameters associated with insulin resistance in obese, hypertensive patients. *Mangifera indica* L has been reported multiple biological activities such as antioxidant, anti-inflammatory, antitumor, antimicrobial (Muruganandan, 2005). Citral is a mixture of cis and trans isomers. It is the main component of lemon grass (*Cymbopogan citratus*) oil and found in all citrus fruits and possesses, antimicrobial, antioxidant, anti-inflammatory activities. Citral was found to possess anticancer effect against prostate gland tumour in various strains of rats (Carbajal, 1989). Epi-catechin isolated from *Pterocarpus marsupim* which was shown to possess preventive as well as restorative properties of b -cells (Si, 2011). Glycerrhizin is the active constituents of *Glycerrhiza glabra* have been reported to increase insulin level and improves glucose tolerance (Saxena, 2005). Curcumin, a principal curcuminoid extracted from turmeric has anticancer, anti-inflammatory, anti-obesity, and anti-diabetic properties (Shehzad et al., 2012). The underlying mechanisms of action seem to involve regulation of redox-sensitive transcription factors, inflammatory cytokines and growth factors. PPAR gamma is one of the most important targets for Curcumin extracted from *Curcuma longa* and 6-gingerol, derived from the root of ginger (*Zingiber officinale* Rosc) (Srivastava et al., 1996). Isoflavones (genistein, daidzein, and glycitein) are present in legumes, grains, and vegetables, but soybeans are the most important source of these polyphenols in human diet (Cederroth and nef, 2009). Quercetin, a flavonol present in apples, onions, scallions, broccoli, apples, and teas, is known to have multiple biological functions, including anti-inflammatory, ant oxidative and anti-mutagenic activities. Quercetin supplementation (10 mg/kg) lessens inflammatory state in the adipose tissue of obese Zucker rats and improves dyslipidaemia, hypertension and hyperinsulinemia. Quercetin also lowers circulating glucose, insulin, triglycerides, and cholesterol levels in mice and a rat fed a calorie-rich diet, and enhances adiponectin expression and secretion (Taesun and Yunyung, 2011). Capsaicinoids and capsinoids, alkaloids primarily found in red hot peppers and sweet peppers, exert pharmacological and physiological actions, including anticancer, anti-inflammatory, antioxidant, and anti-obesity effects (Luo et al., 2011). It has been reported that capsaicinoid consumption increases energy expenditure and lipid oxidation, reduces appetite and energy intake, thus promoting weight loss. Capsaicin also attenuates obesity induced inflammatory responses by reducing tumour necrosis factor-alpha (TNF-α), interleukin (IL)-6, IL- 8, and monocyte chemoattractant protein (MCP)-1 levels (Choi et al., 2011), while enhancing adiponectin levels, important for insulin response (Kang et al., 2011). Several plant-derived flavonoids, apart from possessing their common antioxidant activity, have been reported to inhibit aldose reductase activity and impart beneficial action in diabetic complications (Thielecke and Boschmann, 2009). Recently, Lim et al. (2001) have identified butane as the most promising antioxidant and aldose reductase inhibitor for prevention and treatment of diabetic complications. Further, there is an increasing body of literature indicating that specific phytochemicals have discrete actions on kinase-mediated intracellular signalling processes that are disrupted in patients with chronic diseases. Tan et al. (2008) demonstrated that an extract from bitter melon (*Momordica charantia*) had agonist activity for AMP activated protein kinase (AMPK) in 3T3-L1 adipocytes and enhanced glucose disposal in insulin-resistant mice. In a retrospective sub-analysis of participants with higher cardiovascular disease risk, Lerman et al. (2010) found that the additional phytochemical supplementation was needed in order to adequately reduce multiple biomarkers associated with cardiovascular disease in this high-risk population. Compelling data on type 2 diabetes (T2D) treatment suggest that multiple targeting of the previous metabolic pathways is an acceptable, though not yet fully developed approach to reversing T2D. Pharmacological interference of these targets with anti-diabetic agents has undesirable side effects. Due to the richness and complexity of the compounds in plants, herbal therapy has always been thought to act on multiple targets in the human body. Even one single compound can have multiple targets, the multiple targets associated with anti-diabetic herbal medicine make clinical trials complicated, but such an approach is more likely to lead to an eventual cure for T2D. The stem of *Tinospora cordifolia* is widely used in the therapy of diabetes by regulating the blood glucose in traditional folk medicine of India. It has been reported to mediate its anti-diabetic potential through mitigating oxidative stress (OS), promoting insulin secretion and also by inhibiting gluconeogenesis and glycogenolysis, thereby regulating blood glucose. Alkaloids, tannins, cardiac glycosides, flavonoids, saponins, and steroids as the major phytoconstituents of *Tinospora cordifolia* have been reported to play anti-diabetic role. The isoquinoline alkaloid rich fraction from stem, including, palmatine, jatrorrhizine, and magnoflorine have been reported for insulin-mimicking and insulin-releasing effect both in vitro and in vivo. Oral treatments of root extracts have been reported to regulate blood glucose levels, enhance insulin secretion and suppress OS markers. The root extract has been reported to decrease the levels of glycosylated hemoglobin, plasma thiobarbituric acid reactive substances, hydroperoxides, ceruloplasmin and vitamin E diabetic rats.

Ensete spp. belongs to the family musaceae has wide therapeutic uses and is used in various treatments like diabetes, kidney stone, leucorrhoea and dysuria. In future it can be drug preparation for treating human diseases.

Ensete superbum is used in psychotic disorders, diabetes mellitus, stomach pain, dysuria and leucorrhoea. On phytochemical analysis it was found that the plant contains glycosides, flavonoids, alkaloids, carbohydrates, phenols and tannins.

Young leaves and long slender stem tops of the *Coccinia grandis* are cooked and eaten as a potherb or added to soups. Young and tender green fruits are used either as raw salads or cooked and added to curries. The hypoglycaemic (blood lowering) effect of *Coccinia grandis* on diabetic patients is very important and its leaves are used as medicine. *Coccinia grandis* (L.) Voigt. is a climbing perennial herb distributed almost all over the world. The leaves of the plant possess ant diabetic, anti-inflammatory, antipyretic, analgesic, antispasmodic, antimicrobial, cathartic and expectorant activities. Every part of this plant is valuable in medicine and various preparations have been mentioned in indigenous system of medicine for various skin diseases, bronchitis and Unani systems of medicine for ring worm, psoriasis, small pox, scabies and other itchy skin eruptions and ulcers. The plant is used in decoction for gonnorhoeae, diabetes and also useful in dropsical condition, pyelitis, cystitis, strangury, snake bite, urinary gravel and calculi. The leaves of *Apama siliquosa* Lam. used for the treatment of diabetes mellitus. The objective of this research work was to find out the tribal medicines for diabetics among Mannans in Kerala.

3. Materials and methods

3.1 Plant materials
The present study included plant species which were *Tinospora cordifolia*, *Ensete superbum*, *Coccinia grandis* wild and *Apama siliquosa* Lam.

3.2 Sample collection
An Ethanobotanical survey was conducted among the tribal peoples (community of Mannankudy) of Kumily Panchayath, Peermade taluk, Idukki district of Kerala state through a personal interview to investigate the medicinal plants used in the treatment of Diabetes. Then plant materials were collected locally and identified taxonomically and stored. These plants parts were used for the purpose of their phytochemical analysis and tests for checking the glucose level. Fresh and tender leaves of selected plants were used for phytochemical analysis.

3.3 Preparation of extracts
The plant parts were taken and air dried for few days under shade. The plant parts then crushed into powder and stored in polythene bags for use. The extract of each

sample was prepared by soxhlet extraction method using ethanol as solvent. The extract was kept in bottle without allowing the passage of sunlight and the extract was filtered and diluted with water before use.

3.4 Phytochemical screening

Chemical tests were carried out on the aqueous extract and on the powdered specimens using standard procedures to identify the constituents as described by Sofowara (1993), Trease and Evans (1989) and Harborne (1973).

3.4.1 Test for Tannins

Two ml of plant extract was taken in a test tube and two ml of water and few drops of five percentage ferric chloride was added and observed for brownish green or a blue-black colouration indicate the presence of tannins.

3.4.2 Test for Flavanoids

Five ml of dilute ammonia solution were added to a portion of the plant extract followed by addition of concentrated H_2SO_4. A yellow colouration observed in each extract indicated the presence of flavonoids. The yellow colouration disappeared on standing.

3.4.3 Test for Terpenoids

Five ml of each extract was mixed in two ml of chloroform, and concentrated H_2SO_4 (three ml) was carefully added to form a layer. A reddish brown colouration of the inter face was formed to show positive results for the presence of terpenoids.

3.4.4 Test for Saponins

Ten ml of the extract was mixed with five ml of distilled water and shaken vigorously for a stable persistent froth. The frothing was mixed with three drops of olive oil and shaken vigorously, then observed for the formation of emulsion.

3.4.5 Test for Steroids

Two ml of extract was taken in a test tube and two ml chloroform and two ml of conc.H_2SO_4 was added. Formation of reddish brown ring at the junction shows the presence of steroids.

3.4.6 Test for Phlobatannins

Deposition of a red precipitate when an extract of each plant sample was boiled with one percentage aqueous hydrochloric acid was taken as evidence for the presence of phlobatannins.

3.7 Statistical analysis

The survey results were analyzed and descriptive statistics were done using SPSS 12.0 (SPSS Inc., an IBM Company, Chicago, USA) and graphs were generated using Sigma Plot 7 (Systat Software Inc., Chicago, USA).

Figure 1. *Tinospora cordifolia*; a) plant climbing to a natural support, b) cut stem parts, c) and d) bottom parts of stem.

Figure 2. *Ensete superbum*; a) fruit with seed, b) fruit opened showing seeds, c) dried fruits d) developing flower, e) complete plant with flower. Photo courtesy: Wikipedia.

Figure 3. *Coccinia grandis*; a) plant in natural habitat, b) collected stem and leaves, c) mature fruits d) leaves and fruits.

Figure 4. *Apama siliquosa* Lam. a) whole plant, b) plant showing leaves.

Figure 5. Dried and powdered samples a) *Tinospora cordifolia*, b) *Ensete superbum*, c) *Coccinia grandis*, d) *Apama siliquosa* Lam.

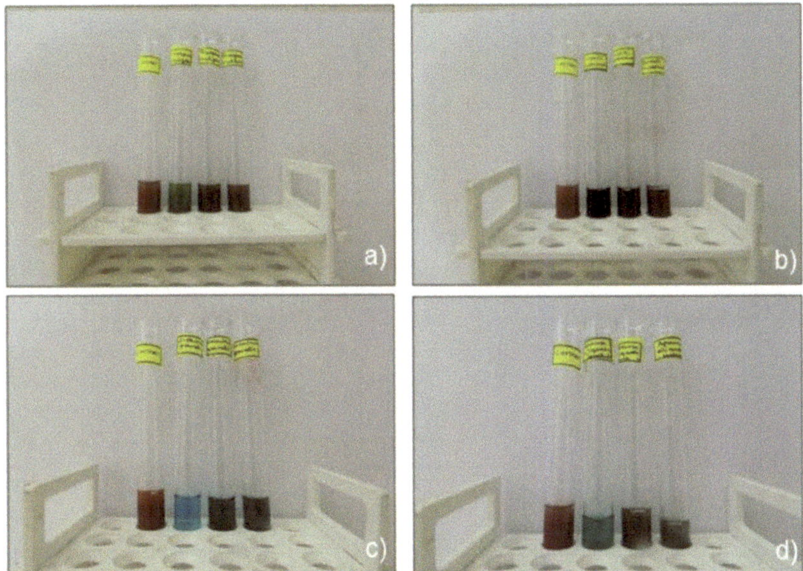

Figure 6. Glucose level tests a) *Tinospora cordifolia*, b) *Ensete superbum*, c) *Coccinia grandis*, d) *Apama siliquosa* Lam.

Table 1. Preliminary phytochemical analysis of screened medicinal plant species.

Sl.No.	Phytoconstituents	*Tinospora cordifolia* (Amurthu)	*Ensete superbum* (Kalluvazha)	*Coccinia grandis* (Kattukoval)	*Apama siliquosa* Lam. (Alpam)
1	Tannins	-	+	+	-
2	Flavonoids	+	-	+	-
3	Saponins	+	+	+	-
4	Steroids / phytosterol	+	+	+	+
5	Phlobatannins	-	+	-	-
6	Carbohydrates	+	+	+	+
7	Glycosides	+	+	+	-
8	Coumarins	-	-	-	-
9	Alkaloids	+	+	+	+
10	Protiens	-	-	-	-
11	Emodins	-	-	-	-
12	Anthraquinones	-	-	+	-
13	Anthocyanins	+	-	+	-
14	Leucoanthocyanins	-	-	-	-

+ = indicates presence of phytochemicals
- = indicates absence of phytochemicals.

Table 2. Changes in the glucose level determined by Anthrone method.

Sl. No	Sample	Glucose level Initial mg/100ml	Glucose Level Final mg/100ml
1	*Tinospora cordifolia*	10	3
2	*Ensete superbum*	10	4
3	*Coccinia grandis*	10	2.5
4	*Apama siliquosa* Lam.	10	5

Table 3. Changes in the glucose level determined by Benedict's method.

Sl. No	Sample	Glucose level Initial mg/100 ml	Glucose Level Final mg/100 ml
1	*Tinospora cordifolia*	10	3.01
2	*Ensete superbum*	10	3.88
3	*Coccinia grandis*	10	2.40
4	*Apama siliquosa* Lam.	10	5

4. Results and discussion

Phytochemical screening of the ethanolic extract of selected herbal plants revealed the presence of compounds like Tannins, saponins, alkaloids, steroids, phlobatannins, carbohydrate, terpenes, glycosides, flavonoids and anthraquinones.

Five kinds of chemical constituents including flavanoids, steroids, carbohydrates, glycosides and alkaloids were isolated from the stem and bark extract of *Tinospora cordifolia*. *Ensete superbum* contains tannins, saponins, alkaloids, steroids, phlobatannins, carbohydrate and glycosides in their seed. Tannins, saponins, alkaloids, steroids, carbohydrate, glycosides and flavonoids were identified from the leaves of *Coccinia grandis*. Extract from the stem and bark of *Apama siliquosa* Lam. shows the presence of phytosterols, carbohydrate and alkaloids.

The experiment to analyse the conversion effect of extract on sugar shows that, *Coccinia grandis* are more effective than the other plants. Following the *Coccinia grandis*, *Tinospora cordifolia*, *Ensete superbum* and *Apama siliquosa* Lam. shows the effectiveness in decreasing order.

Studies shows that *Tinospora cardifolia* (willd) stem extract decreases blood glucose level through glucose metabolism and has an inhibitory effect on adrenaline-induced hyperglycemia(Prince and Menon, 1999, Singh et al., 2003) According to Rahman et al. (2015), extracts of *Coccinia indica* wild reduces blood glucose and glycosylated hemoglobin content. It lowers blood glucose by depressing its synthesis, depression of glucose 6-phosphatase and fructose1,6, bisphosphatase and enhancing glucose oxidation pathway through activation of glucose 6-phosphate dehydrogenase. *Tinospora cordifolia* and *Coccinia grandis* contains a large amount of flavanoids. Flavanoids have the potential to reduce the glucose content in the blood. Alkaloids exert a wide range of antidiabetic activities through different mechanisms. Singh et al. (2003) reported that berberine is isolated from the stems and roots of *Tinospora cordifolia* (Willd) Miers is known to have potent hypoglycemic activity (Singh, et al., 2003). Flavonoids represent a beneficial group of naturally occurring compounds with hypoglycemic potentials. These are widely distributed in plant kingdom and exhibit characteristic pharmacological properties.

Anthocyanins, may also prevent T2D and obesity. Anthocyanins from different sources have been shown to affect glucose absorption and insulin level, secretion, action and lipid metabolism by in vitro and in vivo methods. Many in vitro studies suggest that the anthocyanins may decrease the intestinal absorption of glucose by retarding the release of glucose during digestion. Saponins are bioactive compounds present naturally in many plants and known to possess potent hypoglycemic activity.

All the four plants studied here have a potential for ant diabetes activity. Among them *Coccinia grandis* was more effective than the others because it contains tannins, flavonoids, saponins, phytosterol, glycosides alkaloids and anthocyanins. All these phytochemicals have an ability to reduce the sugar level in the blood. *Tinospora cordifolia* lack tannins and *Ensete superbum* devoid of flavonoids and anthocyanins, comes second and third respectively. Among the four plants *Apama siliquosa* Lam. shows little activity for reducing sugar level.

Screening of four selected medicinal plants clearly reveals that the maximum classes of phytoconstituents are present in *Coccinia grandis* extract as compared to other three selected plant extracts. Hence, the above plant extract could be explored for its highest therapeutic efficacy by pharmaceutical companies in order to develop safe drugs for various ailments. The other three studied plants are of equal importance due to the presence of most of the tested major phytoconstituents. Since these plants have been used in the treatment of different ailments, the medicinal roles of these plants could be related to such identified bioactive compounds. The quantitative analysis of these phytocompounds will be an interesting area for further study. Efforts should be geared up to exploit the biomedical applications of these screened plants due to the presence of certain class of phytocompounds for their full utilization.

5. Conclusions

The present study was conducted to investigate the use of traditional medicinal plants in treatment by the people of a tribal community, Mannans, in Idukki district of Kerala state. It was observed that several traditional medicinal plants were used by the local people for the treatment of diabetes. The plant parts such as seed, leaves, stem and bark were used by the local people. The plant materials prepared as decoction, infusion, aqueous extracts in milk or honey were used for the treatment of diabetes. The plants that are being used by the local people of the study area have been isolated from plants for the treatment of diabetes. The efficacy of these ethnomedicinal plants needs to be subjected to pharmacological validation. Ethnobotanical survey is most useful for scientists, research scholars and scientific companies for further studies on isolation and identification of active compounds that can be formulated into antidiabetic drugs.

Acknowledgements
The authors are grateful for the cooperation of the management of Mar Augsthinose college for necessary support. Technical assistance from Binoy A Mulanthra is also acknowledged.

References

Abo K.A, Fred-Jaiyesimi AA, Jaiyesimi AEA (2008). Ethno botanical studies of medicinal plants used in the management of diabetes mellitus in South Western Nigeria. *Journal of Ethno pharmacology* **115**(1) 67-71.

Ahmad F, Khan MM, Rastogi AK, Chaubey M, Kidwai JR (1999). Effect of (-)epicatechin on cAMP content, insulin release and conversion of proinsulin to insulin in immature and mature rat islets in vitro. *Indian Journal of Experimental Biology* **29**(6) 516-520.

Andrade-Cetto A, Heinrich M (2005). Mexican plants with hypoglycaemic effect used in the treatment of diabetes. *Journal of Ethno pharmacology* 99(3) 325-348.

Bailey CJ, Day C (1989). Traditional plant medicines as treatments for diabetes. *Diabetes Care* 12(8) 553-564.

Bever BO (1980). Oral hypoglycaemic plants in West Africa. *Journal of Ethno-pharmacology* **2**(2) 119-127.

Carbajal D, Casaco A, Arruzazabala L, Gonzalez R, Tolon Z (1989). *Journal of Ethnopharmacology* 25(1) 103-107.

Cederroth CR, Nef S (2009). Soy, phytoestrogens and metabolism: a review. *Molecular and Cellular Endocrinology,* **304**(1) 30-42.

Chattopadhyaya R, Pathak D, Jindal DP (1996). Antihyperlipidemic agents: A review. *Indian Drugs* **33**(3) 85- 98.

Chauhan A, Sharma PK, Srivastava P, Kumar N, Dudhe R (2010). Plants having potential anti-diabetic activity: a review. *Der Pharm. Lett* **2**(3) 369-387.

Chehade J, Mooradian A (2000). A rational approach to drug therapy of type2 diabetes mellitus. *Drugs* **60**(1) 95-113.

Choi SE, Kim TH, Yi SA, Hwang YC, Hwang WS, Choe SJ, Han SJ, Kim HJ, Kim DJ, Kang Y, Lee KW (2011). Capsaicin attenuates palmitate-induced expression of macrophage inflammatory protein 1 and interleukin 8 by increasing palmitate oxidation and reducing c-Jun activation in THP-1 (human acute monocytic leukemia cell) cells. *Nutritional Research* 31(6) 468-478.

WHO (1999). Definition, diagnosis and classification of diabetes mellitus and its complications.Report of a WHO Consultation. Part 1: Diagnosis and classification of diabetes mellitus. Geneva: World Health Organization (WHO/NCD/NCS/99.2.)

Donga JJ, Surani VS, Sailor GU, Chauhan SP, Seth AK (2010). A systematic review on natural medicine used for therapy of Diabetes mellitus of some Indian medicinal plants. *An International Journal of Pharmaceutical Sciences* 1(2) 148.

Enwere OO, Salako BL, Falade CO (2006). Prescription and Cost Consideration at a Diabetic Clinic in Ibadan, Nigeria: A Report. *Annals of Ibadan Post graduate Medicine* 4(2) 35–39.

Gill GV, Mbanya JC, Ramaiya KL, Tesfaye S (2009). .A sub-Saharan African perspective of diabetes. *Diabetologia* 52(1) 8-16.

Grover JK, Yadav S, Vats VJ (2002). Medicinal plants of India with anti-diabetic potential. *Journal of Ethnopharmacology* 81(1) 81-100.

Harborne JB (1973). *Phytochemical methods*, Chapman and Hall, Ltd. London.

Kang JH, Tsuyoshi G, Le Ngoc H, Kim HM, Tu TH, Noh HJ, Kim CS, Choe SY, Kawada T, Yoo H (2011). *Journal of Medicinal Food* 14(3) 310-315.

Lacueva CA, Remon AM, Llorach R, Sarda UM, Khan N, Blanch GC, Ros ZR, Ribalta RM, Raventos RML (2011). In: Rosa LA, Parilla EA, Aguilar GAG Fruit and Vegetable Phytochemicals, Vol 1 (53-88). Wiley- Blackwel, USA .

Lerman R, Minich D, Darland G, Lamb J, Chang J, Hsi A (2010). Subjects with elevated LDL cholesterol and metabolic syndrome benefit from supplementation with soy protein, phytosterols, hops rho iso-alpha acids, and *Acacia nilotica* proanthocyanidins. *Journal of Clinical Lipidology.* 4(1) 59-68.

Luo XJ, Peng J, Li YJ (2011). Recent advances in the study on capsaicinoids and capsinoids. *European Journal of Pharmacology* 650(1) 1-7.

Moller DE (2001). New drug targets for type2 diabetes and the metabolic syndrome. *Nature* 414(6865) 821–827.

Mukherjee PK, Mukherjee K, Maiti K (2006). Leads from Indian medicinal plants with hypoglycemic potentials. *Journal of Ethnopharmacology* 106(1) 1–28.

Muruganandan S, Srinivasan K, Gupta S, Gupta PK, Lal J (2005). Effect of mangiferin on hyperglycemia and atherogenicity in streptozotocin diabetic rats. *Journal of Ethnopharmacology* 97(1) 497- 501.

Patel DK, Kumar R, Laloo D, Hemalatha S (2012). Natural medicines from plant source used for therapy of diabetes mellitus: an overview of its pharmacological aspects. *Asian Pacific Journal of Tropical Disease* **2**(3) 239-250.

Perez YY, Jiménez-Ferrer E, Zamilpa A, Hernández-Valencia M, Alarcon-Aguilar FJ, Tortoriello J, Román-Ramos R (2007). Effect of a polyphenol-rich extract from Aloe vera gel on experimentally induced insulin resistance in mice. *American Journal of Chinese Medicine* **35**(6) 1037–1046.

Prince PS, Menon VP (1999). Antioxidant activity of *Tinospora cordifolia* roots in experimental diabetes. *Journal of Ethnopharmacology* **65**(3) 277-281.

Rahman MA, Sarker J, Akter S, Al Mamun A, Azad MA, Mohiuddin M, Akter S, Sarwar MS (2015). Comparative evaluation of antidiabetic activity of crude methanolic extract of leaves, fruits, roots and aerial parts of *Coccinia grandis*. *Journal of Plant Sciences* **2**(6-1) 19-23.

Rao BN (2003). Bioactive phytochemicals in Indian foods and their potential in health promotion and disease prevention. *Asia Pacific Journal of clinical Nutrition* **12**(1) 9-22.

Lai PK, Roy J (2004). Antimicrobial and chemopreventive properties of herbs and spices. *Current Medicinal Chemistry* **11**(11) 1451-1460.

Saxena S. (2005). *Glycyrrhiza glabra*: Medicine over the millennium. *Natural Product Radiance* **4**(5) 358- 367.

Shaw JE, Sicree RA, Zimmet PZ (2010). Global estimates of the prevalence of diabetes for 2010 and 2030. *Diabetes Research and Clinical Practice* **87**(1) 4-14.

Shehzad A, Khan S, Sup Lee Y (2012). Curcumin molecular targets in obesity and obesity-related cancers. *Future Oncology* **8**(2) 179-190.

Shokeen P, Anand P, Murali YK, Tandon V (2008). Antidiabetic activity of 50% ethanolic extract of *Ricinus communis* and its purified fractions. *Food and Cheical. Toxicology* **46**(11) 3458-3466.

Si H, Fu Z, Babu PV, Zhen W, Leroith T, Meaney MP, Voelker KA, Jia Z, Grange RW, Liu D (2011). Dietary epicatechin promotes survival of obese diabetic mice and *Drosophila melanogaster Journal of Nutrition* **141**(6) 1095-1100.

Singh SS, Pandey SC, Srivastava S, Gupta VS, Patro B (2003). Chemistry and medicinal properties of *Tinospora cordifolia* (Guduchi). *Indian Journal of Pharmacology* **3**(2) 83-91.

Sofowara A (1993). Medicinal plants and Traditional medicine in Africa. Spectrum Books Ltd, Ibadan, Nigeria.

Son MJ, Rico CW, Nam SH, Kang MY (2011). Effect of Oryzanol and Ferulic Acid on the Glucose Metabolism of Mice Fed with a High-Fat Diet. *Journal of Food Science* **76**(1) 7-10.

Engelhard YN, Gazer B, Paran E (2006). Natural antioxidants from tomato extract reduce blood pressure in patients with grade-1 hypertension: a double-blind, placebo-controlled pilot study. *American Heart Journal* **151**(1) 100-106.

Srivastava SK, Awasthi S, Wang C, Bhatnagar A, Awasthi YC, Ansari MH (1996). Attenuation of 4-hydroxynonenal-induced cataractogenesis in rat lens by butylated hydroxytoluene. *Current Eye Research* **15**(7) 749-754.

Subodh S (2010). Production and productivity of medicinal and aromatic plants in Mughal India: A study of contemporary texts. *Asian Agri History* **15**(1) 3-12.

Taesun P, Yunyung K (2011). *Antiobesity Drug Discovery and Development* **1**(1) 14-20.

Tan MJ, Ye JM, Turner N, Ke CQ, Tang CP (2008). Antidiabetic activities of triterpenoids isolated from bitter melon associated with activation of the AMPK pathway. *Chemistry & Biology* **15**(3) 263-73.

Tang GY, Li XJ, Zhang HY (2008). Antidiabetic components contained in vegetables and legumes *Molecules* **13**(5) 1189-1194.

Thielecke F, Boschmann M (2009). The potential role of green tea catechins in the prevention of the metabolic syndrome—a review. *Phytochemistry* **70**(1) 11-23

Trease GE, Evans WC (1989). Pharmacognsy.11th edn.BrailliarTiridel Can. MacMillan publishers.

Venkatesh S, Reddy GD, Reddy BM, Ramesh M, AppaRao AVN (2003). Antihyperglycemic activity of Caralluma attenuata. *Fitoterapia* **74**(1) 274-279.

Surh YJ (2002). Anti-tumor promoting potential of selected spice ingredients with antioxidative and anti-inflammatory activities: a short review. *Food Chemistry and Toxicology* **40**(8) 1091-1097.

Venkatesh S, Reddy GD, Reddy BM, Ramesh M, AppaRao AVN (2003). Antihyperglycemic activity of *Caramulla attenuata*. *Fitoterapia* **74**(3) 274-279.

Wallace TM, Matthews DR (2000). Poor glycaemic control in type2 diabetes: a conspiracy of disease, suboptimal therapy and attitude. *Quarterly Journal of Medicine* **93**(6) 369–374.

WHO (2008). Traditional Medicine.⟨http://www.who.int/mediacentre/factsheets/fs134/en/⟩.

WHO (2013). In: World Health Organization (Ed.), WHO Traditional Medicine Strategy 2014–2023. WHO Press, Geneva, Switzerland.

Yeh GY, Eisenberg DM, Kaptchuk TJ, Phillips RS (2003). Systematic review of herbs and dietary supplements for glycaemic control in diabetes. *Diabetes Care* **26**(4) 1277–1294.

Yusuff K, Obe O, Joseph B (2008). Adherence to anti-diabetic drug therapy and self management practices among type-2 diabetics in Nigeria. *Pharmacy World & Science* **30**(6) 876–883.

Zimmet P, Alberti KGMM, Shaw J (2001). Global and societal implications of the diabetes epidemic. *Nature* **414**(6865) 782–787.

YOUR KNOWLEDGE HAS VALUE

- We will publish your bachelor's and master's thesis, essays and papers

- Your own eBook and book - sold worldwide in all relevant shops

- Earn money with each sale

Upload your text at www.GRIN.com and publish for free